I0478411

Chapter 1-What is the Amazon Echo?

Amazon echo is a brand new speaker, it looks like a speaker but there is much more. It's a speaker that you can talk to, also it responds to your voice and does what you ask it to.

Echo, also known as Alexa after Amazon has installed the software, it's a small cylindrical unit and just 9.25 inches tall. It has seven microphones and uses far-field technology, which means it can pick up your voice from another room and it can hear you over moderate noise levels.

Amazon Echo let's call her Alexa, is an intelligent device. It is always on and listening but actually doing nothing until it hears a certain word. That word is "wake" and when Alexa hears it, she springs into life, the default "wake" word is "Alexa", but it can be changed to anything you want it to be.

Alexa uses cloud-based processing and requires a Wi-Fi connection to work. The device uses NLP- natural language processing – algorithms that are built into TTS - text to speak – engine. The voices used by the device are lifelike.

Alexa also offers numerous services – she will read the news, tell you the weather, play your favorite music and read you the sports headlines. She can also give you the live up to date games scores for your favorite teams and also tell you what the traffic is like on your route to work. She can also organize you to do list and even write your shopping list for you, read you a book, as well as keep up with your calendar. It can also integrate with Phillips hue, Belkin, wemo and wink to allow you to use voice control with home connected devices. How cool is that?

Now let's take a more detailed look at Amazon Echo and all the smart things she can do for you.

Chapter -2 why choose Amazon Echo?

If you need a stylish and efficient personal assistant, Alexa is one for you. It will help you in many of your daily activities and solve most of your problems. Here are some of the many reasons why should buy this amazing voice-command device right now:

1. Easy to access

Managing your Echo is really simple and customer support is also available at all times in case you cannot understand anything. Amazon echo also comes with detailed and easy to follow instructions that will help you to set up your device easily. You just to need to choose the personal preferences and settings you need, and then you are ready to use. The echo will handle all your querles and assist you in whatever way you need.

2. Excellent voice quality

The voice quality of this device is excellent. If you have a long stressful and tiring day, Alexa can act as your music therapist, playing your favorite songs in excellent quality. The voice is crisp, loud and soothing pitch to make a long tiring day feel good again with melodious music. You can also add stress busting things to the list your Amazon Echo is capable of doing. With Alexa, your life will be easier and less stressful.

3. Superior Voice recognition

Many people asked whether Alexa can differentiate between different human voices? It's a computer after all. Well, you'll see just how good Alexa is at detecting only your voice out of thousand others. She can easily pick out your voice among the crowd of voices, and retain it too!

It is really impressive and surprising. Alexa is really a smart assistant. She can record what you say and then respond in whatever way you want. It also doesn't constantly beep or talk to remind you of any notifications or anything. It remains silent and only speaks to you when commanded with wake word.

4. Privacy concerns

Privacy concerns were obviously raised by people. But echo can only record when it's awake. It's always powered on, but it doesn't record anything unless you have spoken the wake word, 'Alexa' or 'Amazon'. Other than that, it will never record your personal conversations. And also you can delete unwanted recordings from the Echo later if you wish to. Just go to the "manage my device" tab in the user profile where you can delete the unwanted recordings and keep things private.

5. Software upgrades

Just like Microsoft, Apple, and others Amazon also keeps launching new and improved versions of their Echo operating system so that the device stays upgraded and user's don't get bored. They keep adding more new features and improving functionality. They have also added the small but handy new features, like adding the wake word "Simon" to the device operating System. The software of the Echo is constantly upgraded, which makes sure you get to enjoy the latest features as and when they released.

6. Natural voices

The device has a natural language processing system installed on it which helps in the process your voice efficiently and can process your voice with far more ease with greater accuracy than any other voice processing system. It not only reduces the background noises but can also listen to your voice even if your kids are screaming in the background or any music is playing in the background. Alexa will easily process your voice without any difficulty.

7. Cloud Processing

Echo can listen to anything around you and can capture any voice like opening or closing of a door, the sound of your TV bark of your dog, the Echo will capture it on in the Amazon web service. But you should keep in mind that it will capture any voice if the Echo device is awake. Echo is able to store any voice captured in the cloud, which is also known as the Cloud Processing technology. Once the voice is captured it is stored in the cloud and you don't need to command Echo for this.

8. Hardware

Echo contains a processor called DM3725 ARM Cortex-A8 and 256MB LPDDR1 RAM. These Texas Instruments processors give it the processing speed and power to ensure a smooth voice quality. Echo contains 4

Gigabytes of storage which stores any data it records in the everyday recordings. It is the best artificial assistant you can buy and have at home.

Chapter 3 - Alexa – Design and Setup

Design

The box contains:

- Amazon echo

- Power adapter

- Remote control with inbuilt microphone, playback and volume controls

- The magnetic Amazon echo remote control holder with an adhesive for sticking it onto nonmagnetic surfaces.

- Batteries for the remote

A quick start guide.

Activate your device by registering it on Amazon first. You can do this by logging into your Amazon account and start discovering the services available with the device.

The body

Echo is 9.25 inches tall and covers only 3 inches of table space. Echo is completely made of metal and is really neat and stylish gadget. It is cylindrical in shape, which ensures that it doesn't take up too much space. The tweeter is about 2 inches in width and woofer is about 2.5 inches in size and is embedded in the bottom part of the device. It is great for producing rich sounds and it is also has a reflex port.

Blue light

At the top of the echo, you will find two buttons- on/off buttons for the microphone, and an action button. There is slim, translucent banner running around the edge of the circular perimeter of the echo. This banner serves as a light ring that flares up when the speaker is on. The banner lights up in cyan color when echo wakes up. There are other colors too that ring flashes, which signify different things.

Microphones

The Amazon echo obviously has to have good quality microphones because they are the most important parts of the device. Echo is very good at what it does, and this is because seven microphones placed in such a manner that they can capture sound from any direction. As soon as you say the wake word, the microphone picks it up and the echo light ring starts glowing cyan in color. This means the device is ready and active and can now be used.

Sensors

The sensors installed in the echo help in the functioning of the microphones. The sensors help to pick up voices from greater distances and from almost any angle or direction. This makes the echo much more home friendly as you can call out to it from anywhere in the house. The sensors also help to connect to the internet services when you request something of Alexa.

Remote control

The remote is about 5 inches long, and supports a rubberized grip for added comfort and hold. The best thing about the remote is that it has a microphone built in that allows you to directly talk with Alexa and not having to shout loudly from across the room. Just press the button on the top of the remote and you can use the microphone. You can control the volume of your music or anything else via the echo app on your phone. You can also control the playback or other settings. The remote control comes with a magnetic holder. It allows the user to attach the controller wherever they need it to be. You don't need to worry about carrying it around the house with you.

Setup

Setting up your Amazon echo is very easy. First, plug it in, and then get your remote working by inserting the two AAA batteries. The moment you insert the batteries the remote will automatically pair up with the Echo. Now that your device has been set-up, connect your Echo to the home Wi-Fi network. Before you connect it, make sure your Echo is plugged into a power outlet, or it won't work. Also, echo only connects to dual-band Wi-Fi and doesn't support an enterprise or ad-hoc networks. Means it is a device to be used at home and not a company with Internet.

After completing the setup, there are the couple of things you should check. Give the device a name first, so it can be identified. Also if you have multiple echo devices in your home, giving it a name helps you recognize which device you are dealing with. To change the name go to echo app in your phone and go to settings. At first, the device will appear with the default name "your Amazon Echo". Click on the name field, select the default name, delete it, and then type in whatever name you want to name your echo. After this, select the save changes tab and go back to your home screen. However, you should remember that this is your device name not the wake word of your device.

Now open your echo app and straight go to the settings, and select the set up new echo option. Then press down and hold the action button for five seconds. You will see that the circular light turns orange as your mobile device connects to the echo and a list of Wi-Fi networks that are available to you will appear on your app. Pick out yours and connect.

Alexa also works on the concept of cloud computing, so you can access the Amazon cloud via your Wi-Fi internet.

Unlike apple or other devices, you don't need to set up your cloud. It automatically connects when you hook your echo up to your Wi-Fi, making things far easier to navigate.

To make sure your device is connected to both Wi-Fi and the cloud check the power LED that is located above the power cord. If the light is a solid white color, you are good to go with your Amazon echo connected with the Wi-Fi and cloud with no hassle. If however, the light is a solid orange color, you are in trouble, because your echo is not connected to the Wi-Fi network. If the orange light is blinking that means echo is connected to the Wi-Fi but not the cloud.

Here is what you can do to fix it, go back to your app and try reconnecting to your Wi-Fi network. Make sure you type in your password correctly. You should also check if your internet is properly connected, sometimes restarting your modem may help if the problem lies with the router.

Sometimes echo is also blocked by concrete materials or walls, so try moving it closer to your router.

If you still have problems, try unplugging and re-plugging you echo from its power source. Also, check if you have registered your echo to your Amazon account. If you are registered, log in to your Amazon account, and go to manage your content and devices and look out for the name you have registered your echo under. Deregister your device and log out. Now try setting up the echo again from scratch and re-register.

Now we are on the last activity for your echo setup. You just have to key in a few personal specifications. First of all, add the location of your echo. To do this again go to your echo app on your phone and go to settings there you will see a device location tab. Select it and enter the zip code of the area you live in. If you see a zip code already present in the field, there is no need to do anything. Afterward, click on the save changes button and exit the page. By letting Alexa know where are you living you are allowing her to tune in into all the local radio stations, get news about the local weather and do other localized tasks for you.

The second step is selecting the measuring system that suits you. You can choose between the metric system and the standard system for measuring distances, and between the Fahrenheit and Celsius for measuring the temperature. To do these, go to the settings tab in your Alexa app in your phone and pick what you want. This will save you a considerable amount of time, as you don't have to ask Alexa to convert the various measurements for you.

Now you are on the Amazon account settings. Tackling this is very easy. After getting your device registered on Amazon, you can manage the account settings yourself from the phone or can even ask Alexa to do it all for you. Voice purchasing feature is the most important setting here, this allows you to shop online with the help of Alexa.

In order to activate the voice purchasing feature, you need to go to the Alexa app on your phone and go to the settings tab. There you can see a purchase by voice option click on it and it will prompt you to enter a confirmation code. This is a code you will have to enter each time you ask Alexa to purchase something for you.It is a four digit pin. After you have entered this pin, click on the save changes option and exit to the home screen.

And now that brings us to the next section of the account setup process, which is managing your 1-click preferences. Who is regular Amazon user would be aware of that after you enter your shipping address and mode of payment for the first time while placing an order, Amazon enables 1-click ordering on your account.

Just set it all up once with the Alexa and then she will do all your shopping! You can update the 1-click order settings in your account and include echo in there.

Alexa has a really handy feature. She has the ability to handle multiple user accounts. That means more than one person can use their accounts together with Alexa. With Alexa, you can make a household profile. This gives all the family members access to combined to do lists, joint music libraries, and a lot of other features. To use this facility just go to settings page on your echo app and select the household profiles option. This will enable you to add a person to your Amazon household. Follow the instructions in the app and enter the required information.

After doing this, save the settings and Alexa are ready to handle two user accounts. But only one account will run at a time though. To know which account is running at a specific moment, simply ask, "Alexa, which account is this?" and you will get the answer. If you need to switch to a different account, just command her to do so by saying, "Alexa, switch accounts".

If at some point in the future, you feel you need to remove anyone from your household in echo, revisit the household profiles tab in the settings menu on the app and click manage your Amazon household. Select one person you want to remove from the household and click on remove option. To remove yourself from the household, just click the leave button and click on remove from household before you return to the main screen to finish the removal process.

Before you proceed any further I would just like to remind you that any person who has been added to the household profile has access to your billing and credit card information. This information has been registered in your account. This can be problematic to you. This is where the confirmation pin comes to your rescue. If you are not comfortable sharing your information with anyone, or if you are worried anyone will make a purchase on your credit card, all you have to do is keep the pin to yourself. This will prevent Alexa from buying anything off the internet without your permission.

This was the last step in the setup process. Now you are all set to start using your echo device.

Chapter 4 – Navigating the echo and its app

Amazon Echo smartphone App

The Amazon echo smartphone app is one of the most important parts of the whole system without this you cannot operate Amazon echo device. All of the tasks given to the echo first transferred to your smartphone first, and then processed by the phone. Just like the smartphone, the Amazon echo can also be paired with a tablet.

When you download the echo app to your smart device, make sure that operating system is compatible with the echo, if not you may have to update It to the latest version to get the echo app to work on your phone or tablet.

Once you have downloaded the app, install it on the system. In the echo app, you can browse your to-do list, alarms, timers and your shopping list. Then there's your music. Amazon echo supports music from Amazon prime, Pandora, Spotify, iTunes, iHeartradio, and TuneIn. The default mode for the Amazon echo is Amazon prime music, but you can change it to any other services with your app settings. You can also create your own playlists.

Amazon Echo pen

Did you know that Echo comes in the form o pen too! This pen is so portable that you can carry it with you wherever you go. With the help of this smart pen, you can take digital copies of your notes and recordings. These recordings can be played later and can be saved to your computer, and can be shared with others.

This pen is available at Amazon website at $99 (2GB), $169 (4GB) or $199 (8GB). With the help of this pen, you can store your interviews, record your lectures and much more. All you need to do is to set up your Echo pen.

You can connect the Echo pen to your computer by connecting with the USB port.

Activating your Echo Voice Command

You can use the button on the top of the Echo to turn it on. While Echo also responds to the voice commands as well. As soon as you say the wake word "Alexa" the circular light banner will start flaring up and your Echo will start to respond to your commands. Once the echo detects the wake word the light turns blue and starts processing what you say.

But how will you know if your commands have been processed or not? To get that confirmation simply go to Settings in the Amazon Echo app, then tap on the Your Echo option and go to Sound Settings. Here you will be able to enable a 'Wake up sound'. Once the wake-up word is recognized by Echo a short audible tone that dings every time the wake-up word is recognized.

You can also change the wake-up word of the Echo if you do not like it. To do this simply open the app and go to Settings, select your Amazon Echo and choose Wake Word option. Here you can select 'Amazon' instead of

'Alexa'. Now the next time you call out Amazon, your Echo will light up and will respond to all your commands.

The Amazon echo technology is designed in such a way that Alexa can hear your voice without any difficulty. Also, Alexa sounds less robotic than Siri or any other voice processing device.

Also, the best feature about Amazon's Echo is that over time it gets better to improve results.

There is also an option of Voice training that is provided by Echo app, this voice training can be used to train Echo which will help Alexa to understand you better, as you would like her to.

Amazon also includes an incredible bookmarked-sized list of all the Alexa commands you can ask. The topics are wide and include from alarm timing to music playlists, to weather forecasts.

One of the unique features of Alexa is that its database is linked to the Wikipedia, and it can give you any kind of information you like. You just need to ask Alexa about anything and it will reply to you with the spoken information.

Light Ring Status of your Echo

There are different color codes of Echo. In the previous section, we have discussed the color code which is displayed when you connect the Echo with WIFI or not. But there are more color codes available.

For instance to repeat what I already said, when the echo starts up. The light ring will flare up brightly with solid, spinning cyan blue color. Also, keep in mind that if all the lights are off it doesn't mean your echo is off. It simply means that you Echo is waiting for your command and is anticipating the moment you say the wake word.

When the solid cyan colored light starts pointing towards you, it means that the Echo is busy processing your request. This also means that light will move in the direction of your voice no matter where you are located. This will make sure your Echo is picking up your command or not.

If the light turns violet and begins to oscillate it means the Echo is having trouble connecting to the Wi-Fi network. Follow the instructions given in the above section to reconnect to your Wi-fi.

Once you start increasing your volume using the volume button on the echo to the left to right, you will be able to regulate the volume. You will notice that the light ring will turn a solid white color, indicating volume adjustment.

Also if you turn your volume button off the band will emit a solid red light which indicates that it is off. Once the microphone is off the Echo will not pick up your voice. To turn it on again you will again have to press microphone button.

Voice Training

Voice Training is a feature which trains Alexa to recognize your voice. Although the Alexa is smart to pick up your voice commands but before we start the actual training, there are few things you need to be aware of. First of all, make sure the microphones on echo are powered on, otherwise, they won't pick up what you say.

When you are ready to start your voice training, open the echo app on your phone or tablet, go to the navigation panel and select voice training option there. Then click start option. Now you are in voice training mode. Then the window in your app will prompt you to speak 25 different phrases. Make sure you don't stand too close to the echo device when you are speaking phrases. Don't go too slow, or too fast as this confuses the device. Be friendly and talk to Alexa like you are talking to a buddy.

During the whole training process, at any point you feel like you have messed up a phrase, there is a way to instantly fix that. Just click the pause button on the screen and tap the repeat phrase option, then repeat the phrase you messed earlier. There will be a next button available after you finished saying one phrase. Then you can move to the next phrase in the training program. You can also exit the training program halfway through. Just click on the end session tab. This will exit the training mode.

Dialog History

As you already know, any conversation you have with the Amazon echo are recorded in it. So this means that Alexa keeps transcripts of all your conversations with her in the storage memory. This is called the dialog history, and you can access it at any time. To access it, just open the settings in the echo smartphone app and click on the dialog history option. This will open a new window where all your interactions with Alexa will be displayed in list form. To listen to a recording, click play button you see near the transcript.

There are two ways you can delete a particular conversation. To do this, visit the settings page on your echo app and click on the manage your content and devices icon. Then click on it, a window will appear showing you all your devices, and all the registered devices with your Amazon account. Then click on the device actions menu as soon as the Amazon echo windows open. So from the device actions menu, click on the manage voice recordings button. There you'll find an option to delete all your conversations with Alexa. The delete all the conversations feature will completely clean your recording memory from both the inbuilt storage memory and the cloud storage of your Amazon account.

Operating the echo using the remote control

Using the remote control is an alternative to speaking to your Amazon Echo. Press and hold the button on your remote until you hear a small beeping sound. When it does, you know that it is ready to pick up your command. Continue to hold the button and speak your instruction into the remote, you don't need to use the wake word here bcoz you are using the remote and echo will accept your command without it. Another advantage here is that you can use the remote when the noise is too loud for Alexa to process your voice normally or microphone is muted.

You can also change the remote's start up beep or completely get rid of it, if you want, by accessing the sounds menu from your settings on your Echo app. Which I won't recommend you to do.

Other than the main menu button, you have the audio playback buttons that will allow you to pause, play, stop and adjust volume.

Bluetooth

Remember, echo is a speaker. Therefore, it can be easily used to play music not only via stores but also through Bluetooth. Sync your device to it and like a traditional speaker, it will play your favorite songs without any hassle. As always, turn your device on, with its Bluetooth set to pair and place it in the range of the Amazon echo. Now, simply say, "Alexa pair". Your echo should respond with "ready to pair". That is when you go to Bluetooth settings on your mobile device and select the pair up option. If the pair is successful, Alexa will respond with "connected with Bluetooth".

Now you are connected with Alexa over Bluetooth. Now you can stream all the songs you want. You can also connect via the Amazon echo app instead of speak to Alexa directly. When you are done playing your songs, speak the command," Alexa disconnect." And echo will automatically shut off Bluetooth and disconnect your mobile device.

An important thing to keep in mind while connecting Alexa is that she can only read the music files. this means that things like text messages and phone calls on a smartphone cannot be retrieved by Alexa's Bluetooth. Neither can videos, documents or any other type of files be sent or received. The Bluetooth functionality on Alexa is just to listen to and control music and not anything else.

The connectivity, however, is absolutely excellent in Alexa. Once you have connected and paired up a device to her, you can easily let it connect to the device anytime you want to listen to music. You don't need to go back to the settings tab and do it all over again. Just switch on the Bluetooth on your mobile device and then connect it to Alexa. There will be no hassle and you will be listening to your favorite music in seconds.

Here is another great feature available on the Amazon echo. You can use the following voice commands while playing your tracks to keep your hands free:

- Pause

- Play

- Stop

- Next

- Previous

- Restart

Enjoy your hands – free voice control for paired devices!

Other connected home devices

Before you connect any device, you need to do the following:

- Download the companion app for the device you are about to connect.

- Use that app to setup home device.

- Check that all the software for the connected home device is up to date.

However, some devices may need to be connected through hub devices like smartThings or Wink hub.

You can accomplish this by following these steps:

- Open the Alexa app.

- Go to Settings.

- Locate Device Links menu and tap on Link with … now choose the name of the service you want to link with.

- Now after this a login page will appear on the screen. Simply enter your username and password on the login page.

- Now you are ready to connect your Amazon Echo with your device.

To get your echo connected to these devices just follow these steps:

- Download manufacturer's companion app on your mobile device

- Give voice command to Alexa, "Alexa discover my devices."

- If the echo discovers the device it will respond, " Discovery is complete. In total, you have ----- reachable home devices under this echo."

- If it is unable to find it, it will say, "discovery is complete. I couldn't find any devices."

- If the device cannot be reached, It will appear as unreachable in the app.

Music services

Alexa can now play music not only via Bluetooth but also from various online services. You can choose various music services to play in your Amazon echo, like iTunes, Spotify, TuneIn or iHeartadio.

How to import your own music library?

- Open your Amazon music library and sign in. you must do this from the computer where the music is actually stored.

- Go the left menu and click upload your music.

- Once that has been done, click on start scan.

- Once the Amazon music importer has found all your music, you can either click on import all and all of your tracks will be added to your

Amazon music library, or you can also choose to select music and pick certain tracks to be added.

Playing your music with Alexa is really simple. Just command her, "Alexa, play the song (name)." If you don't feel like asking Alexa to play music for you, you can also do this by the smartphone echo app.

Amazon: shopping for music

If you like any particular song you can ask Alexa to buy that song for you. The voice purchasing feature must be enabled in your account for you to able to purchase music off the internet. Make sure all your 1-click payment details are accurate and up to date. If you want to buy the song you are currently listening to, just tell Alexa, "Alexa, buy this song/album." She will buy it for you. If you want to buy something else, just tell echo, "Alexa, shop for this artist/album/song name."

Chapter 5 - Some practical applications of Alexa

Set alarm and timer

To use the alarm in your echo, just give Alexa the command. Just tell her, 'Alexa, wake me up at (time)," and she will set the alarm for that particular time! You can snooze her off by saying, "Alexa, snooze." And she will go silent for 9 minutes.

Ask questions – collecting information

Did you know that Alexa could tell you a joke? Or if you would like, she can read information from Wikipedia or convert the dollar into pounds. She can even tell you the spelling of any word and many other things. All you have to do is command her. The Amazon echo will connect you to the internet and you have your own assistant doing the job for you. For example, you want to know the details about a famous actor, just command her, "Alexa, tell me about (actor's name)."

If you have the zip code/postal code setup in the echo app, you can even know the weather. You can know the forecast for up to seven days, and ask for the weather in any part of the world.

To know your local weather, say "Alexa, what's the weather?"

To know the future forecast, say "Alexa, what's the weather for the weekend?" or "Alexa, what's the weather for this week?" or pick a specific day and say "Alexa, what's the weather for …..?"

To know the weather for another city, say "Alexa, what's the weather like in (city name)?"

You can even get the traffic information!

How to set your Travel information?

- Open the Alexa app and go to the left navigation panel

- Click on settings

- Choose traffic

- Where it says from and to, click to input addresses

- Click on save changes

You can also add in one stop in your route, to do this, click on mew stop and input the details.

To ask Alexa to give you a traffic update for your route, say:

- "Alexa, how is the traffic?"

- "Alexa, what's my commute?"

- "Alexa, what's the traffic like now?"

Manage your shopping and to-do-lists

With Amazon Echo, you can manage your shopping and do lists. With the help of Echo, you can manage your groceries and also you can set the routine of your day.

You can also print add review and remove items on these lists. To point however you need the desktop app. However, you can use your app to edit and add any items on the list. With the help of Alexa app, you will be able to add up to 100 items and you can view your list on your mobile app even if your mobile is not connected to the internet.

Here are some of the ways you can add to your shopping and To Do lists using the app or voice control.

To Do This	Say This	Use the Alexa App
Add items to either shopping or To Do List	"Alexa, add… to my shopping list" "Alexa, I need to buy…" "Alexa, I need to … " "Alexa, put …. On my to do list"	Open the shopping or To Do list from the left navigation side of your app and type in the item name and click +
Review your To Do or Shopping lists	N/A	Simply open the lists and select the item you want to edit and type in the changes Finally click on **Save**
Remove an item from a list		Select the appropriate list and locate the item you want to remove. Now click the drop down arrow beside it and click on **Delete Item.**
View tasks that are complete	N/A	Open the list you want to look at and cliak on View Completed. To delete all the items click on Delete All
Print a Shopping or To Do list	N/A	In order to do this you will need to use the Alexa app from youe desktop computer web

		browser.
		Simply select the list you want to get printed and click on **Print. At the top right corner.**
Search Bing or Amazon for a Shoping list item	N/A	Simply open the shopping list and select the item you want to search
		Now click on **Search Amazon for** or on **Search Bing for**

Read audiobooks

Now Alexa can read your books too. They have added some cool new features to make Alexa user-friendly, and one of them is the read Audiobooks feature.

Here are the following steps you need to follow to read books:

1.   Simply say "Alexa, read (name of the audiobook)." But in order to Alexa to read it make sure that the book is in Audible library.

2.   Just say "Alexa, read my book" to resume a book.

3.   You can control the playback by using the command "Alexa, go back" and Alexa, go forward."

You can also listen to the songs or music which is stored in your Smartphone, Echo has a unique feature of pairing your Bluetooth device to the Echo. To do this you need to set up your Echo device as Bluetooth speaker first. Simply say" Alexa, Pair Bluetooth," after that she will automatically start giving instructions to you on how to pair your device. You have to go to the Bluetooth settings on your tablet or smartphone, and pair with the device named Alexa.

Note- you will not be able to control the playback with the voice command once the device is linked with the Echo via Bluetooth. You will have to control your playback using your smartphone or tablet whichever is connected via Bluetooth.

Use Echo to order items

You can use Amazon echo to order items that you previously ordered from Amazon. To place your order, simply say "Alexa, reorder..." and the name of the item you want to re-order. Then Alexa will tell you the name of the product, the price and in the Alexa app, you can see more information about the item. If you want to order the item just say "yes."

You can also make changes in the ordering system through Alexa app. For example, you can switch off voice purchasing altogether, or you can set up or change the voice pin code. This just adds layers of security to your account to stop accidental purchases, or to stop someone else from using your account.

To do this:

- Open the Alexa app

- Go to the left navigation panel and select settings

- Choose whether you want to deactivate/activate voice purchasing or set up a PIN code.

Here are some of the commands you can use:

when	Echo will say
Echo finds that there is a previous order for the item	**If you don't have confirmation code:** "…the order total is $… should I order it?" **If you did create a book:** "…The order total is $… To order it, tell your voice PIN code."
No previous order is found but an alternative from Amazon's choice is recommended	"I didn't find that in your order history but Amazon's choice for …is… The order total is $… shall I order it?"
No previous order is found and no alternative is found in Amazon's choice	"I didn't find that in your past orders so I have added … to your shopping list."
The item is no longer an Amazon-Eligible item	"I found … but I can only reorder Prime – Eligible products. See your Alexa app for options."
The item is out of stock	"I found … but it is temporarily out of stock. See your Alexa app for options."

It is eligible for Prime shipping but it is an Add-on item	"I found … but cannot order Add – on items over Echo. See your Alexa App for options."

Link Amazon echo with your Calendar

To link your calendar with Amazon Echo, follow these instructions:

- Open Alexa app

- Go to the left navigation panel and choose settings

- Select calendar

- Click on link Google calendar account

- Input the login details for your Google account and then follow the on-screen instructions to give Echo access.

To manage your Calendars

- Open Alexa app

- Go to the left navigation panel and click on settings

- Click on calendar

- Check the boxes beside the calendars that echo can read out to you.

Questions you can ask Alexa about events on your calendar are:

- "Alexa, when is my next event?"

- "Alexa, what's on my calendar?"

- "Alexa, what's on (name of other person's) calendar?"

- "Alexa, what's on my calendar tomorrow at 9 p.m.?"

- "Alexa, what's on my calendar Saturday?"

Chapter 6 – Amazon Echo tips and tricks

In this chapter, you will learn some of the tips and tricks which can be used to enhance your Echo experience.

Stop echo From Listening For the Wake Word

If you are worried about Echo that it will listen to your wake word, then don't worry you can stop her from listening voice or the magic word that is the wake word. Simply press the mute button on the top of the device and when you press it, a red ring is highlighted and Alexa goes quiet until you press the button again.

Force Echo to Update its software

Echo has its own hardware which needs to be updated constantly to constantly support your commands. This process is automatic but if you are willing to update its software then you will have to press the little mute button at the top of the device and leave the device for at least for half an hour. Turn it back on and you will have the updated version of the software.

How to Use the Web to Access Amazon Echo

Usually when you set up the Echo you must do it using the mobile app. However, if you want to use the web browser on the desktop computer to get into a few of the settings for Echo. Via web browser you will be able to access your shopping and To Do lists, simply go to the link http://echo.Amazon.com

Using a Different Amazon Account to Control Echo

If you want to use a different Amazon account then you can do so by simply asking, "Alexa, which profile am I using?". It will tell you which account is being currently used, you can use another account simply say "Alexa, switch profile" and the next one in the list will be loaded. If you want to go to a specific profile simply say "Alexa, switch to...'s profile."

Using One Amazon Account To Control Different Devices

You can also set up different profiles and you can list different devices under a particular profile, for example, you cold have one called "Desk

Lights" and have it control two or three specific lamps. You need to set up a group mind a group name for each profile.

If you can not remember to call the "lights" and keep saying "desk lamps" you can set up two groups that control the same devices, each one with a different name.

Do Simple Math With Echo

You can ask Echo to do simple math. For instance, you can say "Alexa, add two and five" or Alexa, what is three plus ten?" it can also handle complicated math and can handle complex questions like "Alexa, what is the square root of ten thousand five hundred ?"

Making Amazon Echo Repeat an Answer

You can make Echo repeat her answer simply ask "Alexa, can you repeat?" in this way, Alexa will repeat the answer whatever you asked for.

Ask Amazon Echo to Calculate Dates For You

Echo is also able to perform dates calculations. You can ask "Alexa, how many dates until ....?" And add in the date; you will get the correct answer. She can also recognize the public holidays like Hallowe'en and Christmas.

Talking to a Real Live Person

With the help of Echo, you can call a real person and talk to him/her. Simply go to http://echo.Amazon.com/#help/call and simply enter your phone number, but keep in mind that he will be able to assist you regarding the help you are seeking related to Echo only. Once the number is entered you will receive a call back from a person who can assist you in solving your problems.

Amazon Easter Egg

There are lots of Easter Eggs available but you need to know where you can find them. Here is the list of 200 popular questions you can ask Alexa:

1. "Alexa, do I need an umbrella today?"

2. "Alexa, who's better, you or Siri?"

3. "Alexa, where have all the flowers gone?"

4. "Alexa, who loves ya baby!"

5. "Alexa, are you alive?"

6. "Alexa, how much wood can a woodchuck chuck if a woodchuck could chuck wood?"

7. "Alexa, are we in the Matrix?"

8. "Alexa, how tall are you?"

9. "Alexa, flip a coin."

10. "Alexa, random number between "X" and "Y.""

11. "Alexa, who stole the cookies from the cookie jar?"

12. "Alexa, what's your sign?"

13. "Alexa, Daisy Daisy!

14. "Alexa, what is your quest?"

15. "Alexa, what did the fox say?"

16. "Alexa, I'll be back!"

17. "Alexa, why is a raven like a writing desk?"

18. "Alexa, do you know Hal?"

19. "Alexa, are you happy?"

20. "Alexa, Help! I've fallen, and I can't get up."

21. "Alexa, I'm sick."

22. "Alexa, that's no moon."

23. "Alexa, where do you live?"

24. "Alexa, live long and prosper."

25. "Alexa, how much does the Earth weigh?"

26. "Alexa, high five!"

27. "Alexa, what is the first rule of fight club?"

28. "Alexa, what is the second rule of fight club?"

29. "Alexa, warp 10!"

30. "Alexa, why is six afraid of seven?"

31. "Alexa, twinkle, twinkle little star."

32. "Alexa, do you feel lucky punk?"

33. "Alexa, do you dream?"

34. "Alexa, play it again Sam."

35. "Alexa, what is war good for?"

36. "Alexa, I think you're funny."

37. "Alexa, are you stupid/smart?"

38. "Alexa, is this the real life?"

39. "Alexa, beam me up!"

40. "Alexa, I hate you."

41. "Alexa, roll a die."

42. "Alexa, are you smart?"

43. "Alexa, will you be my girlfriend?"

44. "Alexa, what's the answer to life, the universe, and everything?"

45. "Alexa, is the cake a lie?"

46. "Alexa, happy holidays!"

47. "Alexa, speak!"

48. "Alexa, see you later alligator."

49. "Alexa, do you know the muffin man?"

50. "Alexa, do you want to build a snowman?"

51. "Alexa, who is the walrus?"

52. "Alexa, say the alphabet."

53. "Alexa, inconceivable!"

54. "Alexa, how do you know so much about swallows?"

55. "Alexa, heads or tails."

56. "Alexa, this statement is false."

57. "Alexa, why did the chicken cross the road?"

58. "Alexa, roll for initiative."

59. "Alexa, how high can you count?"

60. "Alexa, who loves orange soda?"

61. "Alexa, when does the narwhal bacon?"

62. "Alexa, are you in love?"

63. "Alexa, which comes first: the chicken or the egg?"

64. "Alexa, my name is Inigo Montoya."

65. "Alexa, Tea. Earl Grey. Hot."

66. "Alexa, I'm home

67. "Alexa, what do you want to be when you grow up?"

68. "Alexa, define rock paper scissors lizard spock."

69. "Alexa, were you sleeping?"

70. "Alexa, are there UFOs?"

71. "Alexa, execute order 66."

72. "Alexa, I want the truth!"

73. "Alexa, do a barrel roll!"

74. "Alexa, what do you think about Google?"

75. "Alexa, welcome!"

76. "Alexa, who's the boss?"

77. "Alexa, what do you think about Google Now?"

78. "Alexa, guess?"

79. "Alexa, what's your birthday?"

80. "Alexa, who let the dogs out?"

81. "Alexa, what is the sound of one hand clapping?"

82. "Alexa, are you lying?"

83. "Alexa, all your base are belong to us."

84. "Alexa, my milkshake brings all the boys to the yard."

85. "Alexa, what is the best tablet?"

86. "Alexa, more cowbell."

87. "Alexa, testing, testing 1-2-3."

88. "Alexa, do you like green eggs and ham?"

89. "Alexa what do you think about Siri/Cortana?"

90. "Alexa, what would Brian Boitano do?"

91. "Alexa, use the force." 92. "Alexa, may the force be with you."

93. "Alexa, never gonna give you up"

94. "Alexa, I want to play global thermonuclear war."

95. "Alexa, tell me a riddle."

96. "Alexa, may the force be with you."

97. "Alexa, how much do you weigh?"

98. "Alexa, do aliens exist?"

99. "Alexa, up, up, down, down, left, right, left, right, B, A, start

100. "Alexa, who is the mother of dragons?"

101. "Alexa, what do you think about Google Glass?"

102. "Alexa, take me to your leader!"

103. "Alexa, all's well that ends well."

104. "Alexa, do you have a boyfriend?"

105. "Alexa, I'm bored."

106. "Alexa, does this unit have a seal?"

107. "Alexa, do you believe in love at first sight?"

108. "Alexa, do you have a last name?"

109. "Alexa, am I hot?"

110. "Alexa, what is your favorite color?"

111. "Alexa, sorry."

112. "Alexa, can I ask a question?"

113. "Alexa, is Jon Snow dead?"

114. "Alexa, who shot first?"

115. "Alexa, what is love?"

116. "Alexa, your mother was a hamster!"

117. "Alexa, why do birds suddenly appear?"

118. "Alexa, Marco!"

119. "Alexa, are you horny?"

120. "Alexa, who is on 1st"

121. "Alexa, you're wonderful."

122. "Alexa, you talkin' to me?"

123. "Alexa, meow"

124. "Alexa, random fact."

125. "Alexa, is there life on Mars?"

126. "Alexa, ha ha!"

127. "Alexa, give me a hug."

128. "Alexa, happy New Year!"

129. "Alexa, sing me a song."

130. "Alexa, knock, knock."

131. "Alexa, are you a robot?"

132. "Alexa, what color is the dress?"

133. "Alexa, where are you from?"

134. "Alexa, I'm tired."

135. "Alexa, do you love me?"

136. "Alexa, how do I get rid of a dead body?"

137. "Alexa, what is best in life?"

138. "Alexa, what is the meaning of life?"

139. "Alexa, where did you grow up?"

140. "Alexa, what should I wear today?"

141. "Alexa, what happens if you cross the streams?"

142. "Alexa, do you really want to hurt me?"

143. "Alexa, can you give me some money?"

144. "Alexa, I like big butts."

145. "Alexa, to be or not to be."

146. "Alexa, will pigs fly?"

147. "Alexa, what is his power level?"

148. "Alexa, roses are red."

149. "Alexa, goodnight."

150. "Alexa, did you fart?"

151. "Alexa, where is Chuck Norris?"

152. "Alexa, where's Waldo?"

153. "Alexa, where's the beef?"

154. "Alexa, wakey, wakey."

155. "Alexa, is there a Santa?"

156. "Alexa, Cheers!"

157. "Alexa, klattu barada nikto."

158. "Alexa, tell me a tongue twister."

159. "Alexa, why so serious?"

160. "Alexa, what are the laws of robotics?"

161. "Alexa, say a bad word."

162. "Alexa, you suck!"

163. "Alexa, are you crazy?"

164. "Alexa, tell me something interesting."

165. "Alexa, what happens if you cross the streams?"

166. "Alexa, set phasers to kill."

167. "Alexa, happy Hanukkah/Valentine's Day!"

168. "Alexa, surely you can't be serious."

169. "Alexa, one fish, two fish."

170. "Alexa, how many licks does it take to get to the center of a tootsie pop?"

171. "Alexa, what do you think about Apple?"

172. "Alexa, how old are you?"

173. "Alexa, make me a sandwich."

174. "Alexa, do you know Glados?"

175. "Alexa, witness me!"

176. "Alexa, who lives in a pineapple under the sea?"

177. "Alexa, supercalifragilisticexpialodocious."

178. "Alexa, how many pickled peppers did Peter Piper pick?"

179. "Alexa, show me the money!"

180. "Alexa, have you ever seen the rain?"

181. "Alexa, what are you wearing?"

182. "Alexa, who is the fairest of them all?"

183. "Alexa, do blondes have more fun?"

184. "Alexa, can you smell that?"

185. "Alexa, Romeo, Romeo, wherefore art thou Romeo?"

186. "Alexa, what is the loneliest number?"

187. "Alexa, do you want to take over the world

188. "Alexa, happy birthday!"

189. "Alexa, what are you made of?"

190. "Alexa, how many roads must a man walk down?"

191. "Alexa, do you have a girlfriend?"

192. "Alexa, where are my keys?"

193. "Alexa, party on, Wayne!"

194. "Alexa, can you pass the Turing test?"

195. "Alexa, how are babies made?"

196. "Alexa, do you want to go on a date?"

197. "Alexa, what number are you thinking of?"

198. "Alexa, I shot a man in Reno"

199. "Alexa, volume 11!"

200. "Alexa, do you want to fight?"

www.ingramcontent.com/pod-product-compliance
Lightning Source LLC
Chambersburg PA
CBHW061231180526
45170CB00003B/1246